THE ALIENATED PLANET

SAVING LIFE ON OUR MAJESTIC PLANET

FARID POURKHATAI
*also known
as Fred Katai*

Contents

"I conceive that pleasures are to be avoided if greater pains be the consequence."

Michel de Montaigne

I have always been intrigued by life since I was a kid. After years of studying, researching, and teaching, I have come to appreciate the inevitable links amongst science, ecology, technology, economy, and our societies. In this unconventional book, I share with you how we, humans, have ignored or misinterpreted these connections in the past few millennia and as a result have lost touch with science, and the way our natural world is organized and sustained life for billions of years. As I completed studies towards my degrees in biochemistry and biology, I began to notice the common themes occurring amongst the levels of organization in nature [Subatomic Particles, Atoms, Molecules, Macromolecules, Cell Parts/(Organelles), Cells, (Tissues, Organs, Systems), Organisms, Populations, Communities, Ecosystems, Biomes, Biosphere], particularly within the biotic components. We must therefore apply the same themes to our communities. The rule book provided by our generous and gracious host, Mother Nature, contains the survival strategies that ensure the well-being of all participants at various levels of organization.

At the molecular level, I have been fascinated by the complexities of the organic reactions in the cell. However, the chemist in me always doubted the notion of "emergent properties" as the putative (regulatory) processes that define life, clearly span all levels. In my decades of teaching, while attempting to portray a simplified version of the most complex machine, the cell–the basic unit of life, surprisingly, many significant themes of life would readily surface. The prominent themes such as spatiotemporal compartmentalization, simplicity in complexity, unity-diversity, structure-function, nature-nurture, and dormancy govern not only the events at the cellular, but at multiple levels of organization. Hence, The Cell Theory must have an added statement declaring the cell as the basic unit of the biosphere. The objective of our existence must be to fight for and protect the most sacred and complex entity in the universe. The preservation of the basic unit of "us" and the biosphere equate to the preservation of Mother Nature, and is the ultimate duty of all intelligent beings.

These same scientific principles must be applied to our (not so) complex social systems and networks. Science education is indeed the only way humans can come to appreciate the intricacies of life and embrace sustainable living through acknowledging the laws of ecology. As a scientist and educator, I am obliged to put forth this comprehensive scientific ideology which widens the aperture so we can have a larger perspective. In this context, we understand why and how systemic and transformative changes must happen not only to individuals or organizations but also to the foundations of our civilizational systems. A parallel, unifying, global curriculum, for instance must inform and nurture (future) citizens to work collaboratively and innovatively to reconstruct the resilient, sustained communities of tomorrow capable of coexistence with all living things on our planet. It is not surprising that small scale farmers who practice agriculture linked to natural ecosystems by cutting down their use of synthetic fertilizers and pesticides end up experiencing less soil degradation, waste less water and resources, pollute less and are ultimately more profitable, biodiverse, and sustainable.

Environmental movements have been effectively smothered or greenwashed in the past few decades as we have continued to treat anthropogenic (man-made) crises, such as climate change as the cause, not a symptom of our socioeconomic failures. Current environmentalism lacks the necessary scientific foundation to be regarded as a credible movement. Moreover, true scientists lack the essential biological background to lay out a unifying ideology that links thermodynamics to life; Erwin Schrodinger surprisingly attempted it. Here, I share a new perspective on life on our planet. The incredibly complex regulatory processes that ultimately drive the process of natural selection on our, or any life-harboring, planet is indeed dictated by thermodynamic laws. Molecular ecology, a multidisciplinary, and holistic ideology allows the integration of our observations to illustrate why and how we should get from where we are to where we must be. Anthropocentricity or anthropocentricism refers to the notion that humans viewing themselves as the most important entity in the universe, is the obsolete and unscientific belief, even more regressive than geocentricism, as it encompasses nationalism, racism, and sexism. In addition, this procured narcissistic behavior is clearly responsible for our exponentially increasing ecological footprint.

The cost of this book would initiate fundraising towards my goal of setting up an international, non-profit venture that promotes molecular ecoliteracy and invests in other ecological footprint reduction initiatives. Let's call upon the United Nations, the World Bank, and the International Monetary Fund to provide leadership in addressing our crises at a global scale, propelling corporations, and governments towards sustenance. How likely would it be for our existing multilateral and international governing bodies to collaborate and truly address our compounding global challenges? And finally, a call upon the individuals who bear the biggest responsibility of becoming educated to educate, cut down on their footprint-contributing activities to paralyze the mega-corporations of destruction, and demand change.

In this book I discuss the scientific processes that have driven and sustained life for billions of years. This is a bird's eye view of the way our planet functions and the way our actions are dismantling it. Who I think my target audience should be you ask? I think every living human being on this planet who can read, and his or her ecological footprint is not zero should consider reading this book. Also, check out the *"Show Me"* lectures, for further clarification. In this unconventional non-fiction, an imaginary organism, Micro, has found a way to correspond with an imaginary alien, Fred, who lives as a human on planet Earth! I am hoping that a unified (scientific) community's contributions will help to define and quantify the causes of crises; so we can effectively warn policy makers about them. Science calls for us to recognize the interdependence at all levels. I am also optimistic that this book would serve as a manifesto to help humanity figure out how cooperation within and across our communities, from families to international institutions can guide us towards sustenance.

Dear Fred,

I am writing you this letter on behalf of all organisms on our wondrous planet. First and foremost, I must thank you for helping me express the urgent concerns of all living things.

Introduction

Humans, including you and your ancestors, have unknowingly outdone the ecological rules of this gracious planet. The inability and unwillingness of the intelligent mammal to attend to the laws of planet Earth have made us the true, voiceless victims. Humanity must accept that most of the current crises are anthropogenic. Your species must realize the need for a logical, global, and unified strategy to save itself and the planet. The consequences of inaction are becoming exceedingly terrifying. Marginalized humans and we disproportionately suffer the burden of the anthropocentric or anthropogenic schemes. To eradicate these unintelligent cycles, we need a coherent scientific ideology, a feasible strategy, and an effective educational platform, acting as a mass awareness campaign. We certainly hope the multilateral institutions will implement such transformative measures. Humans need to again grasp the secrets of living harmoniously with other organisms and stop the destructive ways that are clearly threatening life on our planet. I have tried to write this message in the most straightforward language possible to act as a liaison between you and us.

Our precious planet has been in a state of distress for the past few millennia – particularly the past few decades. Your exponentially increasing population, resource extraction, production, consumption, and pollution have been placing a tremendous amount of pressure on Mother Nature, our collective life-giver.

Implicit complicities

My Grandma used to tell us generational stories about how and when your ancestors turned against nature. Humans, the descendants of arboreal primates, learned quickly that their only chance of surviving in their new habitats would involve tool making (technology) and division of labor. Apparently, certain groups of the genus Homo or even groups of sapiens started abusing other members within their primitive social constructs. They succeeded in gaining better access to food, resources, space, shelter, and even mating rights, passing on their genes and imprinting their malignant behavior. These seemingly successful, aggressive individuals established colonies that they led; and set up arbitrary, selfish rules ensuring their dominance in these hierarchical classes of subordination. Anthropocentricity was further established, accepted, followed, and passed on once these clans began to practice the process of Artificial Selection some five to ten thousand years ago. The open nature of the ecosystems on Earth clearly has given many humans, even today, the false impression that we live in an available, infinite biosphere! The discovery of fire (combustion) and the ability to control it also played a significant role in this 'myth of invincibility' – a narcissistic, acquired trait that has since passed down the generations. The advances in technology and agriculture helped these anthropocentric clans grow. They easily wiped out nomadic or less aggressive, more harmonious groups and competed ferociously with other empires. Within these last few millennia, Homo sapiens clearly separated themselves from the natural world as they settled and began the process of aggressive farming and civilization.

Population explosion

The need to reproduce had become not only a biological, but a social obligation of humans. Each clan needed to have enough (man)power to fend for territory, fight for its expansion, and tend to the necessities. These clans viewed overbreeding as the means to secure their reign. The growing clans hence went on to overuse the finite resources of their territories to survive. Once depleted, the clan would begin competing fiercely with its neighbors for more resources.

The expanding farms and cities progressively destroyed the land they were built upon. Unfortunately, they also destroyed vast aquatic and terrestrial ecosystems surrounding them. The poor infrastructure that the cities are constructed upon represent unsustainable entities. The farms or cities that replace the natural ecosystems must be maintained through a steady supply of resources and energy. Once again, the seemingly open nature of ecosystems gives the false impression that ridding of your waste out of sight would eliminate it. This downplays the significance of recycling resources. All these factors have been forcing the clans to expand their territories to regain access to depleting or heavily polluted necessities. The hubs of civilization that harbored heavy-polluting and energy-intensive activities of resource mining and processing practiced extensive deforestation. Could the early empires in Africa and Asia have been responsible for the vast stretches of deserts in these continents?

Wars amongst these clans resemble Intraspecific Competition acting as an environmental resistance factor. However, they have not been able to affect the natality rates and instead helped promote the colonial ideologies of these empires.

Thermodynamically speaking, access to energy has allowed humans to sustain their settlements. But, they were also able to satisfy the never-ending apatite of the royals and the ever-growing circle of their supporters and beneficiaries in this micro-period of biological time. The subordinates have continued to accept their roles, convinced that they are chosen to serve and be exploited! In this awful, destructive scheme, many factions were increasingly mistreated. Women and captured enemies, for instance, were treated as slaves or laborers. Many of these racist, sexist, and anthropocentric views continue to this day. This unfortunate scheme has been the foundation of humanity's flawed social constructs in the past few millennia.

From geocentricism to egocentricism

The true turning point in the destruction scheme, however, dates to the Industrial Revolution. This was when your ancestors gained access to a new, liquid fuel that could more easily be transported. The non-solid fossil fuels could be purified and combusted more completely and efficiently. Humans also found many (important) ways to produce various versatile synthetic compounds, collectively referred to as petrochemicals. Once again, while these synthetic chemicals appeared to improve human life, they placed a major burden on the biosphere. The maddening rates of extraction, refining, combustion of these fossil fuels, as well as excessive production of many non-biodegradable synthetic petrochemicals, accelerated the destructive practices of civilization to a (tipping) point of no return. The non-biodegradable nature of these primarily organic petrochemicals renders them persist in the environment. Many anthropogenic compounds, especially non-polar petrochemicals, contain halogens and benzene. These substances tend to bioaccumulate, dismantling many food chains and webs and threatening biodiversity.

Tipping points of no return

Globally reversible, biochemical reactions are positioned naturally at equilibrium as the Earth ultimately represents a closed system. The humans' interference with these delicately balanced systems in a noticeably short time interval clearly justifies the tipping points that we are all accelerating towards.

The accessible and convenient energy sources have opened the door to positive feedback cycles. These have translated into major discoveries and inventions, allowing better access to resources that are not even part of the razor-thin biosphere. The invention of the internal combustion engine, for instance, has clearly played a significant role in these vicious schemes that have progressively magnified the ecological footprint of your species. A transition to a different energy source, such as nuclear fusion, will not change the impact of the feedback loops in these destructive systems. Some influential corporations are opponents of scientific facts. They are continuing to introduce various reform strategies to linger their profit-making schemes of bleeding the voiceless and Mother Nature.

Malignant corporations mismanaged material and energy resources in the post-petroleum era. This clearly confirms the flawed mainstream ideologies that directly or indirectly support colonialism, patriarchy and sexism, racism, and ultimately anthropocentricity. In less than two centuries, the intelligent being has managed to almost use up a natural, nonrenewable resource entirely. This has irreversibly interfered with the global carbon cycle! Should you worry about the unavailability of petrochemicals to future generations? Listing some of the key factors that have increased humans' life expectancy, namely most hygiene methods, and pharmaceuticals may help one answer part of that question.

The 'concentrating' practice of materials interferes with the global material cycles through mining and not recycling. However, the extraction and uncontrolled combustion of fossil fuels for energy purposes have also given rise to an incredible rise in carbon dioxide levels. Carbon dioxide is a reactant in the endothermic process of photosynthesis. The exponential increase in the carbon dioxide concentration abruptly disturbs the equilibrium state of global carbon cycle – one of the many consequences of this is global warming. Since humans have also increased deforestation, these anthropogenic crises are causing compounding damage.

As the local ecosystems have been progressively depleted or polluted, globalization inevitably became the key strategy for large corporations to maintain their diseased economic efforts. Would increased access to energy sources allow them to reach the finite resources at the depths of the oceans, forests, glaciers, or outer planets!? Would rare-metal batteries, genetic engineering, carbon sequestration, and solar dimming solve the problems? Will these reforms act to push the carrying capacity of the planet for your exploding population? Are these the new greenwashing strategies of destruction, moving forward?

With the passing of older indigenous leaders, we sometimes wonder whether there would be any human beings left who truly want to spend time and energy to save Mother Nature? And if so, are those figures waiting for the conditions to worsen before taking proper, precautionary measures? Is there enough time for searching for and analyzing more symptoms to assist with the diagnostics? And then, what would be their most effective revival course of action? These points are rather significant as certain authentic movements have been effectively greenwashed and replaced by superficial reforms which soon restore the destructive status quo.

Integration and differentiation

Scientists seem to dramatically disagree over these issues. The ecologists are the most relevant group. They have population or social demographers that use mathematical and statistical models to predict and project future human population trends and behaviors. Many other ecologists have distanced from experimental sciences even further, leaning towards psychology or spirituality. Conservationists, the ultimate reformists, hope for an unrealistic environmental scheme with preserved wildlife, primarily bird and mammal varieties.

While biologists have recently managed to connect their views to (physical) chemistry, most are specialists. They tend to overlook the unifying themes of life, namely, unity, simplicity in complex systems, and evolution across all levels of organization. Excess detail in many fields of science has obscured the key universal concepts.

Most of the progressive scholars, as a result fall into the category of 'good intent, low impact'. Their research and efforts focus on the consequences (symptoms) of the actions rather than the causes (triggers). Moreover, their incoherent pro-environmental ideologies help them promote or behave in an ecologically responsible fashion. However, their reluctance to acknowledge anthropocentrism and their own enormous ecological footprint translate into actions or strategies with low impact.

Most of the so-called scientists are specialized technologists. They have mastered or developed necessary algorithms and skills of applying scientific principles to help better human lives while serving mega-corporations. And the concept of ethics has never been treated seriously either. Successful industries employ specialized accountants and technologists to exploit loopholes around these loosely defined social or environmental obligations, financially securing the enterprise and its shareholders. This is partially how technological advancements, including inventions and discoveries, get approved for mass production and marketing before their true impacts are investigated.

The pandemics of the previous century clearly showed the inability and incapability of the inferior, anthropocentric leaders and the scientific community of scholars and prize winners. Many are now claiming that they had predicted the occurrence of the Covid-19 pandemic. However, the certainty remains that no one from this (elite) group even would come close to claiming that they had a preventive plan for such calamities. To link this pandemic and many other crises to anthropocentricity lays the grand plan for preventing future pandemics and other environmental crises. It is crucial to reiterate that these novel pathogens emerge through antigenic shifts, involving genetic recombination, and cause zoonotic diseases due to human activities. The Coronavirus causing Covid-19, whether natural or synthetic, is yet another anthropogenic phenomenon. How can these pathogens be compared to invasive species?

Current global leaders, primarily concerned about the economic status of their jurisdiction and their re-appointment are also baffled by, unwilling, or unable to tackle these problems. The unwillingness to confront the reality of the crises may stem from their genuine hope for technology and new energy systems to rescue us all from these predicaments. However, the recent pandemic has clearly proven that placing too much faith in technological capabilities can have serious adverse consequences. Instead of assessing the symptomatic trends, there needs to be a sound analytic, scientific framework. This would enable international policymakers to set guidelines for local governments, organizations, and citizens, informing them about sustainable policies. The need for large-scale transformations toward global sustainability is greater than ever. The continued failure to recognize the finite capacity of the biosphere is propelling us towards the collapse of a perpetual economy, threatening all life forms on the planet. A deliberate attempt to move away from the current malignant economy undoubtedly hinges on 'ecological footprint reduction', guiding humanity towards a regenerative, circular economy.

A global ideology and a unifying curriculum

Education historically has had agricultural and religious ties and loyally has promoted the central messages of anthropocentrism. The exponential and unstoppable rise of the human population can undoubtedly be linked to the spiritual teachings at the places of worship and schools, particularly in the so-called developing parts of the world. These messages continue to be endorsed by the few mega-corporations of food, energy, pharma, and (digital) technology, whose prosperity hinges on increased consumption of their addictive products and services.

Recent advances in (social) media technology have enabled these enterprises to develop and promote schemes that get end users addicted to overconsuming resources. These schemes are based on planned and perceived obsolescence and cost externalization. The supposed new prosperous economy of the developed world deliberately and overtly rewards overconsumption and over-wasting behaviors, undermining the urgency of reproduction/procreation.

The most prevalent Western curricula are intentionally designed to promote the current economic model. It justifies and encourages overextraction, overconsumption, and waste generation. Simultaneously, it promotes sexism, racism, colonialism, and anthropocentricism.

The anti-scientific approaches of this obsolete educational system have also helped raise passive learners and citizens.

The 'just so' stories, Santa Claus, superheroes, and many other cultural renderings tend to turn off the inquisitive minds of young learners. Unfortunately, (social) media and other quasi-educational channels have also contributed to promoting these unnatural and fictional phenomena. The myth of invincibility surprisingly relates to egocentric fantasizing. Humans are entertained to see themselves as the hero of their life stories. The ineffective educational system has clearly failed individuals gain a more realistic view.

How unfortunate is it to see your ecological footprint increase with an increased level of education!? The need for the overhaul of this obsolete system is imminent. An upward counterfactual approach can facilitate the reconstruction of the foundations of a novel system.

Nature's curriculum

Molecular ecoliteracy deciphers the magical lessons we can observe at various levels of organization in nature. Then, it relates them to the way humans must learn to live as part of nature, not against it. There needs to be a progressive, unifying global curriculum based on scientific inquiry. This curriculum would simultaneously cover the analytic elements of the scientific method with the precious components of indigenous education. Such a curriculum will assist intergovernmental systems in engaging individuals, families, communities, and nations to address, reverse, and subsequently mitigate the compounding global challenges. With this model, humanity can restructure and reweave the fabric of its civilization.

Humans must learn the scientific principles and broad physical forces that have been driving life in our biosphere for billions of years. To emulate Mother Nature's recipes, you must design a grand plan aligned with ecological forces.

The thermodynamic laws govern all processes in the universe. So, familiarity with them would help humans appreciate the limits of our closed biosphere. Life on our planet is specifically governed and regulated by the availability of the abiotic factors. These factors exert their impact through basic processes, such as diffusion and osmosis. Every cell is defined by a semipermeable membrane. Every cell houses the key macromolecules of life – carbohydrates, lipids, proteins, and nucleic acids. The biotic components of the ecosystem simultaneously affect life forms through ecological and natural principles.

If we examine any cell in detail, we become aware of its complex, multilayered regulatory biochemical mechanisms. These mechanisms allow the cell to adapt, survive, grow and multiply during the times of plenty; and decelerate and hibernate during scarce times. The environmental resistance factors orchestrate the selection of the fittest organisms. This selection primarily occurs through selective expression of the key regulatory genes that control cell division (cyclins and CdK's) and cellular metabolism (PFk, rubisco, and ATP synthase). These epigenetic mechanisms ultimately give rise to diverse populations, which are also subject to layers of ecological regulation. Can acquired characteristics be inherited? Indeed so. The environment cannot deliberately change the base sequences of the genes. However, the environment can establish heritable, spatiotemporal patterns of chemical markers around segments of the DNA that can dictate gene expression patterns. Can similar mechanisms be responsible for complex interactions such as colonization, offspring rearing, and symbiosis? Has nurture (epigenetics), which entails learning, become the preferred driving force in the process of natural selection over nature (genetics)? Can these same ecological principles be expected to govern any life-supporting planetary system?

We need to reference the incredible dependence of all of life on autotrophs and decomposers. Humans must realize that these organisms define the 'life-generating capacity' of Mother Nature. Humans must learn to appreciate them and recognize the significance of interdependence and biodiversity in preserving and sustaining these majestic organisms. As humans are gradually learning to embrace diversity amongst themselves, they should also be taught the significance of the role biodiversity has played in sustaining life on the planet.

Spatiotemporal compartmentalization is a central theme in the virtuous biochemical cycles of unicellular prokaryotic, eukaryotic cells, and those of multicellular organisms. This compartmentalization clearly indicates that as the degree of interactions amongst the components within the biosphere intensifies, so do the regulatory processes. Stem cells, cellular differentiation and specialization, and the symbiotic interactions within ecosystems are the key illustrations of how nature and nurture interact to maximize the survival chances in a sustained fashion. Reproduction and consumption patterns and restrictions in how these participants operate can clearly paint the constructs of any viable community. Are there correlations between regeneration and reproductive capabilities of organisms? Why are unicellular organisms and autotrophic cells programmed to be immortal – unlike most other heterotrophic ones? Are heterotrophs the selected quasi-parasites that have evolved tighter regulatory, consumptive, and reproductive capabilities?

Hence, the fourth statement of the cell theory must regard the cell as the basic unit of the biosphere! Mother Nature calls for all communities to emulate her grand plan. The following lays out the foundations of this plan, all acquired or learned characteristics:

1) Spatiotemporal compartmentalization requires 'learning to collaborate'. Nurture applies to all levels of organization.

2) Concern for safety: from the cell to the biosphere; Biotherapy: what is unsafe to a living cell is unsafe to all levels! Resourcefulness is practiced at all levels.

3) Interpersonal responsibilities parallel interspecific responsibilities Appreciation of efforts and contributions of other members within the societies and within the biosphere (niche). Justice and reciprocity ensure accountability.

4) Protection of institutional diversity and biological diversity Community-based conservation ensures productivity.

We do not assume that humans are destructive creatures by nature. Contrary to what capitalists believe, the reversal of humans' actions cannot be achieved through total government control. 'The tragedy of the commons' is still taught uncritically as a justification for Feudalistic views or even environmental failures. This philosophy cites that such events can never be averted! But humans were well capable of living sustainably in the past, and many indigenous tribes continue to do so today.

We can observe in any cell or organism the existence of collaborative management systems and governance in well-defined, sustained hubs. This organization translates into resources and benefits being monitored and equally shared. The members or components are trained to survive and innovate to contribute to the well-being of their immediate community. The members are made aware of planetary limits. Trusting relationships with other members need to be established. These 'other members' include the layers of authority, from the pod leaders to the international governing institution. Also, universal, fair, and equitable processes must be in place to deal with the risks, resolve conflicts, and discipline the rule violators.

Exposing the causes of transformation within your current communities must illuminate the reversal pathways, leading to effective mitigation strategies. Once a more harmonious state is achieved, humanity will enact preventative measures of disintegration. Sustained human civilizations of the future will indeed be complex systems like those observed at various levels of organization in nature: specialization of stem cells during embryogenesis and the interactions of neurons in the brain. The collaborative work of living components in these cases show how the interacting pieces are trained to act even without a central control element.

We need a multilateral, intergovernmental endorsement on a scientific ideology and a global educational platform. These two factors will catalyse a unifying, virtuous impactful re-

sponse to help the intelligent beings drastically cut their ecological footprint – rescuing our precious planet.

Yours Truly,

Micro, the Ibex;
pleading to Fred, the alien on the alienated planet

Here is the list of some of the additional lectures referenced in the letter:

https://www.educreations.com/lesson/view/the-story-of-success/56422693/?s=xFGTjj

Molecular Ecology 1 Abiotic Resistance Factors:
https://www.showme.com/sh?h=YKPD9Rg

Molecular Ecology 2 Abiotic Resis Factors 2:
https://www.showme.com/sh?h=DsfRBEu

Molecular Ecology 3 Anthropogenic Pandemics:
https://www.showme.com/sh?h=LO7Smh6

Molecular Ecology 4 Food Energy Waste:
https://www.showme.com/sh?h=KaVAH2W

Molecular Ecology 5 Unsustainable Food Chains:
https://www.showme.com/sh?h=0r22hSS

Molecular Ecology 6 Sustenance:
https://www.showme.com/sh?h=A43naW8

Molecular Ecology 7 Sustenance 2:
https://www.showme.com/sh?h=jdpgsSW

Molecular Ecology 8 Sustenance Education:
https://www.showme.com/sh?h=bNpTBD6

Molecular Ecology 9 Sustenance Education 2
https://www.showme.com/sh?h=oS1qMwi

Molecular Ecology 10 Anthropocentricity
https://www.showme.com/sh?h=fglrvSC

Molecular Ecology 10' Anthropocentricity:
https://www.showme.com/sh?h=7v9qMXQ

Molecular Ecology 10" Global Affairs:
https://www.showme.com/sh?h=zr5mLLc

Molecular Ecology 11 Futures:
https://www.showme.com/sh?h=9DUiCB6

Molecular Ecology 12 Futures 2:
https://www.showme.com/sh?h=SQNaeDg

Molecular Ecology 13 Integration 1:
https://www.showme.com/sh?h=MLKo4Dw
https://www.showme.com/sh?h=OMlJqZU

Molecular Ecology & Ecoliteracy 15
https://www.showme.com/sh?h=zDBq8kC

Molecular Ecology: Intro to Fossil Fuels
https://www.showme.com/sh?h=szHNPcm

The Agenda for Change

A coherent, integrated, and unifying framework to address our compounding and intensifying global challenges

The following are the Key Logical Steps towards Sustenance through the establishment of a harmonious economy:

The Sustainable Development Goals (SDGs) of the United Nations (UN) must be ranked/weighed with:

1) SDG 1 representing Lowering of Ecological Footprint at all levels*. The weight of this goal must be substantially higher than the rest (100) as it encompasses Global Population Reduction (33), Global Resource (Energy, Materials) Mining/ Consumption Reduction (33), and Global Waste Reduction (33).

2) SDG2 must be on Educating towards sustainability (25). A Global/Unifying Curriculum that will help inform masses about the new SDGs.

3) SDG3 must deal with Equal Rights (10) as it aims at Resource Reallocations, Migrations, Border Removals, and Demilitarization.

4) SDG4 must deal with resource-dependent activities (1) pertaining to building new Hospitals, Community Centres, Accessibility/Training Hubs

The UN moves towards taking over the Global Economy (Health Care, Education, Agriculture, Employment, Investments, Accounting, General Services, & Global Policy making/Legal System based on SDGs). i.e. The UN adopts a global educational platform that actively promotes sustenance.

a) The UN presents the new SDGs and its new educational platform to the governments, religious groups, and the public.

b) The new International Taxation system addresses SDG1: In this new paradigm, ecological footprint reducing practices are rewarded, while activities and products that increase them will be exponentially taxed.

Ecological Accounting

This exponential global taxation grid initially targets the production and consumption of non-essential items and activities that leave large relative ecological footprints. Real Estate property taxes, for instance need to be drastically increased; and the inheritance of it must be abolished.

The collected funds get invested in practices that promote sustainability: (Jobs). These practices will set up cascades of change:

— Investment in Individuals, Organizations, Governments with data supporting their footprint reduction measures[1].; Transparent/Accountable, Peaceful, Co-operative in removal of Borders and De-militarization

- Supporting (Non-profit) organizations addressing SDG3

- Minimal Support for (Non-profit) organizations addressing SDG4; i.e. Mining, Infrastructure.

c) The Public mobilizes to acknowledge and implement the new SDGs

- Particularly the marginalised can use Film/Art Industry, & (Social) Media to play a significant (educational) role in the elimination/reversal of positive feedback loops (such as planned or perceived obsolescence) designed to promote overpopulation, overconsumption & unnecessary pollution, and socioeconomic injustices.

- Embrace nature; learn to slow down; dormancy

Resourcefulness: 4R's

Compete and collaborate towards sustainability

Local/Microproduction: Minimalize! Sustainability Pods

Microgreens, microgrids, micropods

Waste reduction, composting

- Adjust diet: Permaculture, Vertical Farming, Aquaponics

Towards the elimination of meat (promote eating at lower trophic levels) and paralyzing the industrial agriculture. Seasonal foods. Preservation methods

- Demand the removal of bullying and symbolic leaders/figures.

- Demand the removal of borders.

- Demand the illegalization of narcotics, alcohol, tobacco, guns.

- Demand demilitarization

- Innovate towards bettering our lives while reducing our footprints.

- The most successful innovations will set up positive feedback loops in virtuous cycles of sustainability.

d) Innovations to better our lives without threatening other (people/organisms). Colleges, Universities and Research Institutions must eliminate research that is:

- Military-related

- Promoting specific corporations (particularly if ecological footprint is high)

Examples: fossil fuel, pharmaceutical, agriculture, digital industries

– Research to quantify key ecological parameters: ecological footprint, sustainability, biodiversity. Introduce new metrics for easier comparative analysis/ Regulation of processes and activities.

– General research on:

– bettering our lives without exceeding our footprints (R&D)

– alternatives to current energy sources and petrochemicals

[1] The governments must adhere to the implementation of SDGs to remain as active/contributing members (Newly elected policy makers cannot undo the commitments)

* Levels in this context refer to "Individuals, Families, Communities, Municipalities, Nations, and the Biosphere"

The numbers in the brackets indicate the weight for each SDG (or subgoal)

A coherent, integrated, and unifying framework to address our compounding and intensifying global challenges

The following are the Key Logical Steps towards Sustenance through the establishment of a harmonious economy:

The Sustainable Development Goals (SDGs) of the United Nations (UN) must be ranked/weighed with:

SDG 1 representing Lowering of Ecological Footprint at all levels*. The weight of this goal must be substantially higher than the rest (100) as it encompasses Global Population Reduction (33), Global Resource (Energy, Materials) Mining/Consumption Reduction (33), and Global Waste Reduction (33)

SDG2 must be on Educating towards sustainability (25). A Global/Unifying Curriculum that will help inform masses about the new SDGs

SDG3 must deal with Equal Rights (10) as it aims at Resource Reallocations, Migrations, Border Removals, and Demilitarisations

SDG4 must deal with resource-dependent activities (1) pertaining to building new Hospitals, Community Centres, Accessibility/Training Hubs

The UN moves towards taking over the Global Economy (Health Care, Education, Agriculture, Employment, Investments, Accounting, General Services, & Global Policy making/Legal System based on SDGs). i.e. The UN adopts a global educational platform that actively promotes sustenance.

1) The UN presents the new SDGs and its new educational platform to the governments, religious groups, and the public!

2) The new International Taxation system addresses SDG1: In this new paradigm, ecological footprint reducing practices are rewarded, while activities and products that increase them will be exponentially taxed.

Ecological Accounting

This exponential global taxation grid initially targets the production and consumption of non-essential items and activities that leave large relative ecological footprints. Real Estate property taxes, for instance need to be drastically increased; and the inheritance of it must be abolished.

The collected funds get invested in practices that promote sustainability: (Jobs). These practices will set up cascades of change:

- Investment in Individuals, Organizations, Governments with data supporting their footprint reduction measures[1].; Transparent/Accountable, Peaceful, Cooperative in removal of Borders and Demilitarization

- Supporting (Non-profit) organizations addressing SDG3

□   Minimal Support for (Non-profit) organizations addressing SDG4

Mining, Infrastructure

[1] The governments must adhere to the implementation of SDGs to remain as active/contributing members (Newly elected policy makers cannot undo the commitments)

3) The Public mobilizes to acknowledge and implement the new SDGs

> Particularly the marginalised can use Film/Art Industry, & (Social) Media to play a significant (educational) role in the elimination/reversal of positive feedback loops (such as planned or perceived obsolescence) designed to promote overpopulation, overconsumption & unnecessary pollution, and socioeconomic injustices.

> Embrace nature; learn to slow down; dormancy

Resourcefulness: 4R's

Compete and collaborate towards sustainability

Local/Microproduction: Minimalize! Sustainability Pods

Microgreens, microgrids, micropods

Waste reduction, composting

> Adjust diet: Permaculture, vertical farming, aquaponics

Towards the elimination of meat (promote eating at lower trophic levels) and paralyzing the industrial agriculture. Seasonal foods. Preservation methods

➢ Demand the removal of bullying and symbolic leaders/figures.

➢ Demand the removal of borders.

➢ Demand the illegalization of narcotics, alcohol, tobacco, guns.

➢ Demand demilitarization

➢ Innovate towards bettering our lives while reducing our footprints.

○ The most successful innovations will set up positive feedback loops in virtuous cycles of sustainability.

4) Innovations to better our lives without threatening other (people/organisms).

Colleges, Universities and Research Institutions must eliminate research that is:

☐ Military-related

☐ Promoting specific corporations (particularly if ecological footprint is high)

Examples: fossil fuel, pharmaceutical, agriculture, digital industries

➢ Research to quantify key ecological parameters: ecological footprint, sustainability, biodiversity. Introduce new metrics for easier comparative analysis/Regulation of processes and activities.

> ➤ General research on:
>
> – bettering our lives without exceeding our footprints (R&D)
>
> – alternatives to current energy sources and petrochemicals

*Levels in this context refer to "Individuals, Families, Communities, Municipalities, Nations, and the Biosphere" The numbers in the brackets indicate the weight for each SDG (or subgoal)

www.ingramcontent.com/pod-product-compliance
Lightning Source LLC
Chambersburg PA
CBHW060509160726
47992CB00003B/1407